180°
90°
90°
(18 ÷ 6 × 5) − 14 ÷ 7
= (3 × 5) − 14 ÷ 7
= (15) − 14 ÷ 7
= 15 − 2
= 13
z
y
x
124
274 | 33976
−274
657
−548
1096
−1096
0
90°
2284 × 4170 =
9,524,
851 | 6
60

For Bharat and your mathematical mind—
thank you for suggesting I write this story
—R. S. R.

To all the children who see magic in numbers
—C. P.

BEACH LANE BOOKS • An imprint of Simon & Schuster Children's Publishing Division • 1230 Avenue of the Americas, New York, New York 10020 • • The text for this book was set in Fairplex Narrow OT. • The illustrations for this book were rendered digitally. • Manufactured in China • 0626 SCP • First Edition • 1 2 3 4 5 6 7 8 9 10 • Library of Congress Cataloging-in-Publication Data • Names: Rajan, Rekha S. author | Prabhat, Chaaya illustrator • Title: The numbers race : how Shakuntala Devi's mathematical mind zoomed around the world / written by Rekha S. Rajan ; illustrated by Chaaya Prabhat. • Other titles: How Shakuntala Devi's mathematical mind zoomed around the world • Description: First edition. | New York : Beach Lane Books, [2026] | Includes bibliographical references. | Audience: Ages 4–8 | Audience: Grades 2–3 | Summary: "A picture book biography of record-breaking Indian mathematician Shakuntala Devi as she takes on the world's toughest math problems"—Provided by publisher. • Identifiers: LCCN 2025046867 (print) | LCCN 2025046868 (ebook) | ISBN 9781665957052 hardcover | ISBN 9781665957069 ebook • Subjects: LCSH: Shakuntala Devi, 1929-2013—Juvenile literature | Women mathematicians—India—Biography—Juvenile literature | Mathematicians—India—Biography—Juvenile literature | LCGFT: Biographies • Classification: LCC QA29.D554 R35 2026 (print) | LCC QA29.D554 (ebook) | DDC 510.92—dc23/eng/20260112 • LC record available at https://lccn.loc.gov/2025046867 • LC ebook record available at https://lccn.loc.gov/2025046868

THE NUMBERS RACE

How Shakuntala Devi's Mathematical Mind Zoomed Around the World

WRITTEN BY
REKHA S. RAJAN

ILLUSTRATED BY
CHAAYA PRABHAT

BEACH LANE BOOKS
New York Amsterdam/Antwerp London Toronto Sydney/Melbourne New Delhi

In a small, dusty, crowded village in South India, little Shakuntala loved to race.

She raced around the streets, chasing 20 auto-rickshaws as the drivers honked their horns.

She raced 45 chickens on the neighbor's farm, quickly scooping one up in her arms as it squawked.

She raced her sister, brother,
3 cousins, and 4 friends around
their family's small hut.

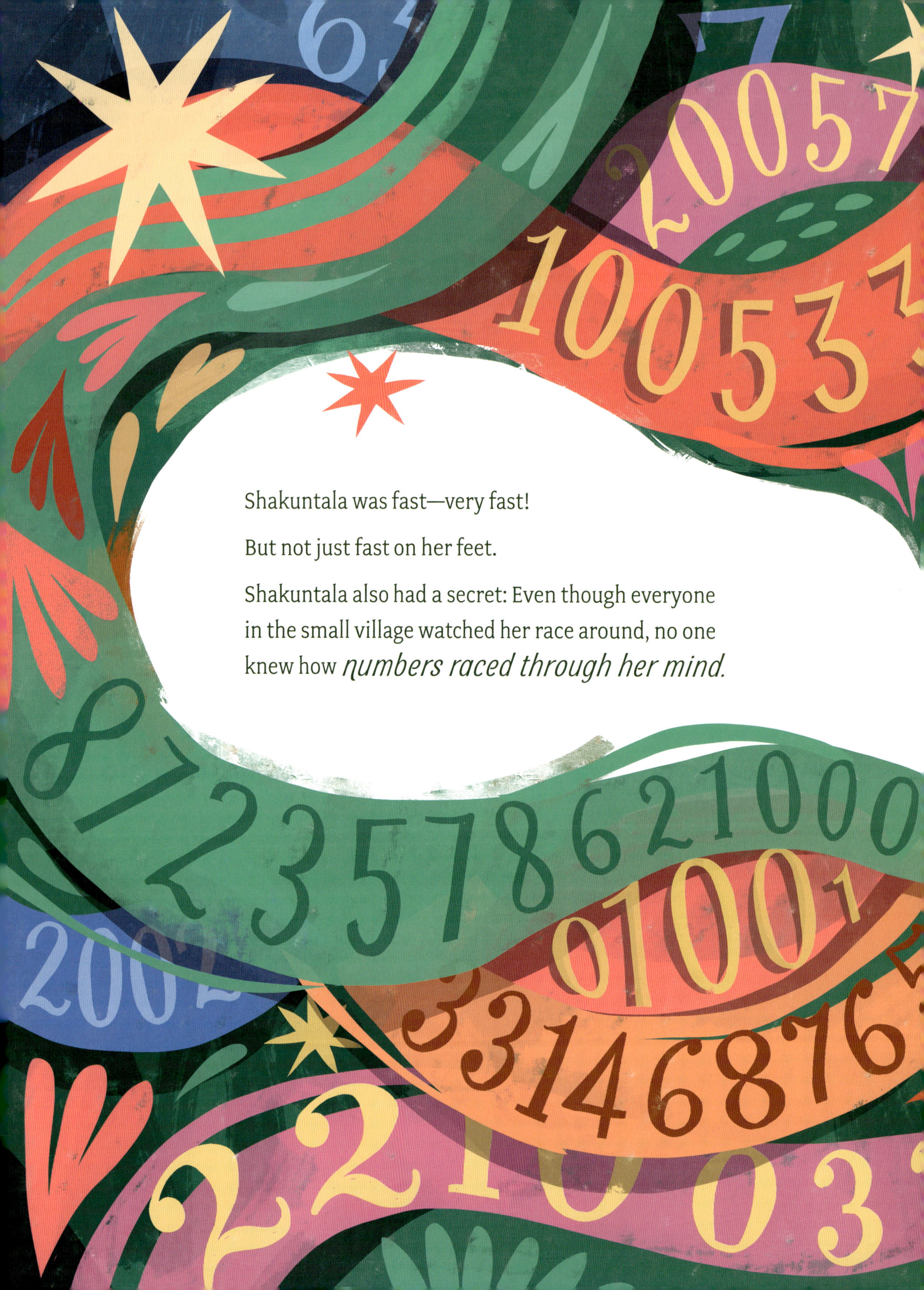

Shakuntala was fast—very fast!

But not just fast on her feet.

Shakuntala also had a secret: Even though everyone in the small village watched her race around, no one knew how *numbers raced through her mind.*

485
2425 ÷ 5
= ?

Until one morning, when her cousin's math homework caught her eye.

"Show me!" Shakuntala said, peeking over his shoulder.

"Dividing numbers is too hard for you," he protested. "You're only three, and I'm twelve."

"The answer is 485," she replied.

Her cousin frowned. "You cheated!"

Shakuntala's thick black braids swung from side to side as she shook her head and said, "I am *not* a cheater."

Shakuntala stomped inside their hut, where Amma poured sizzling, sticky batter to make crispy dosas for lunch.

Shakuntala carefully watched the flips and the folds. "You flipped and folded dosas 111 times!" Shakuntala announced.

Amma looked up, surprised. "I wasn't keeping count."

The evening was Shakuntala's favorite time of day. She held on tightly to Appa's hand as they walked together to his work: a noisy, bustling circus!

Shakuntala had the pattern of the stripes on the big tent memorized.

She counted:

4 ladders!
5 elephants!
10 ropes!
20 horses!

She multiplied how many times the acrobats flipped:
3 times 3 equaled 9 flips!
3 times 10 equaled 30 flips!

When Appa brought out his drums and BANG-BANG-BANGED to tame the curly-maned lions, Shakuntala's brain *ROARED*: 2 lions and 3 BANGS of the drum for each lion equaled 6 beats total!

She galloped like the horses, clapping as they raced around the circus tent—5, 10, 15, 20 laps!

Shakuntala secretly wished she could be a performer too.

Appa noticed how Shakuntala was always counting and memorizing and completing math problems faster than her older cousin.

So, one day, Appa gave her a test.

He gave little Shakuntala a deck of 52 cards and had her mix them up.

Shakuntala watched her hands as she shuffled them into a new pile and handed them back to Appa.

Appa pulled out a card from the middle and quickly hid it behind his back.

"Shakuntala, can you name the card in my hand?" he asked.

Shakuntala smiled. *Numbers raced through her mind.*

"The ten of hearts," she declared, like a magic trick. But it wasn't magic. "I memorized all the cards."

"But you only saw them once! Amazing!" Appa cheered. "Shakuntala, I have something very important to ask you . . ."

That night Shakuntala's family had a big meeting.

Everyone was quiet as Appa and Shakuntala stood up together.

"I have an announcement," he said. "I'm quitting the circus!"

Everyone gasped.

“No!” her brother and sister cried.

“Why?” Amma asked, worried.

“Because,” Shakuntala said, “I am going to perform for the world!”

At six years old, Shakuntala became a performer—just like Appa! They traveled together for thousands of miles all across India.

Professors and students challenged Shakuntala to solve math problems faster than them.

People laughed. *A child could never beat a math expert!*

Shakuntala's stomach tingled. She shook her braids back and forth, getting ready.

The math problems were listed on the blackboard:

2 divided by 3,560

8 divided by 4,528

Numbers raced through Shakuntala's mind.
She shouted out the answers—and she was first!
She beat the adults!

Shakuntala won race after race, performing on bigger and bigger stages, beating more and more difficult mathematical challenges. As a teenager, Shakuntala left India and zipped all across Asia and Europe!

An audience member in London gave her a set of 100 numbers to divide in just 1 minute.

Those were a lot of numbers. The timer would be ticking, and she had to *race the clock.*

Shakuntala straightened the side of her red-and-green ghagra, getting ready. Racing a clock was different from racing people. But she loved math. She loved performing.

Could she do it?

People whispered, *A young girl could never beat that time!*

Shakuntala smiled, looked straight at the audience from left to right, then shouted out her answer before the timer went off.

But wait!

"Shakuntala," the judges said, "you're wrong. Your answer doesn't match the calculator."

Shakuntala frowned. The numbers raced through her head again and again. She turned with a huff and quickly walked away. She knew she had the right answer!

But before she left the room, they called out again—

"Wait! You were correct. The calculator was wrong. The calculator made a mistake!"

Shakuntala flipped the drape of her sari, raised her right eyebrow, and grinned.

Shakuntala loved being onstage.

She loved making the audience laugh.

She loved math.

But no matter how many times she solved math problems, people said it was magic. A trick! She hated when her cousin had called her a cheater as kids. Now, even though she was a grown-up, people *still* called her a cheater.

Shakuntala smiled widely onstage even though she sometimes felt hurt inside. If no one understood how numbers spoke to her, she would show them.

Shakuntala ignored the whispers and giggles around her.

When someone doubted her, she stood taller than the lion tamers in Appa's circus.

When someone laughed at her clothes, she pulled the fabric of her silk sari over her shoulder as her mind *SOARED* higher than the acrobats.

She asked for harder questions—bigger numbers!

People cheered each time she solved a problem.

After each answer Shakuntala would smirk and ask, "Am I correct?" and the audience would roar with laughter. She was always correct.

And soon enough, people stopped questioning and started to believe.

Shakuntala counted three rings before she answered.
It was an important call: She was invited to . . . a race!

But she wasn't racing rickshaws or chickens, her sister, a university audience, or even a clock.

Shakuntala was going to use her mind to *race* against a computer!

The room was packed with people watching. It was her biggest audience ever.

And thousands more people watched from their homes on live television, all around the world.

For the first time in a long time, Shakuntala felt a tingle in her arms and a tickle in her stomach.

People whispered, *A woman could never beat a computer!*

For the first time, she wondered if she *could* win the race.

As the computer whirred to life, Shakuntala sat up straight in her chair and smoothed the silk of her sari.

Next came the hardest math question she had ever heard. The computer gears started turning.

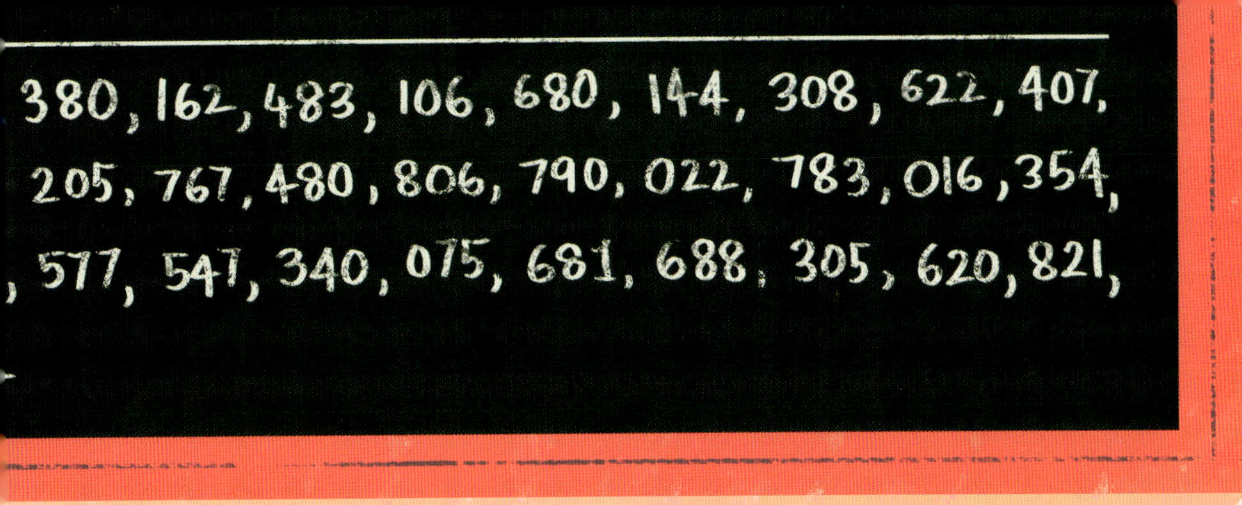

Numbers raced through Shakuntala's mind.

The computer was moving fast, racing through the numbers.

All eyes were on Shakuntala as she suddenly stood up,
smiled, and said the correct answer . . .
before the computer!

The computer kept whirring for another thirty seconds after Shakuntala's mind raced to the answer. The audience roared with applause.

"You're . . . a human computer!" reporters declared. "How do you do it?"

"I'm no computer. I'm a person, and numbers speak to me," Shakuntala said. "Just like you are speaking to me now."

The numbers spoke to Shakuntala, and she spoke back, performing her calculations, counting all the stages and cities, tallying the sizes of the audiences, and sharing her magnificent mathematical mind with the world.

Each time someone asked a math question,
Shakuntala stood up tall and tossed the fabric of
her sari behind her left shoulder,
leaning into the microphone with a smile,

ready to answer,

ready to race the numbers to the finish line.

Photo by TONY AW/*South China Morning Post* via Getty Images

More on Shakuntala Devi

Shakuntala Devi was born on November 4, 1929, in a small village in Bengaluru, India. Her family was too poor to send her to school, and their only income came from her appa's circus. He discovered her natural talent for mathematics while watching her play card games and noticing her ability to memorize numbers and patterns. When he asked Shakuntala to perform math problems in front of small groups, her talent and performance skills started generating a greater source of income for the family. Soon Shakuntala began performing across the country, even though she was only six years old.

Shakuntala continued to perform—and loved performing—all through adolescence and into adulthood. She became known around the world for her strong stage presence, sense of humor, and passion for being onstage. She didn't just solve math problems; she put on a show and made people laugh. And she proved that a woman in a sari could be a master mathematician.

Shakuntala was bullied in many ways throughout her life and career—by children in her village, by adult audience members, and by colleagues at universities. Even after she proved her mathematical talents and abilities over and over again, people still ridiculed and doubted her. Despite people questioning her talent, Shakuntala never gave up! Time and again, she proved she wasn't a cheater or doing magic tricks—she was a mathematician.

Shakuntala's biggest moment came when she accepted a challenge to race against a computer. During that time computers were big, clunky devices that were mainly used to make different types of complex calculations. People thought computers were the only way to calculate large math problems quickly. But Shakuntala proved otherwise.

When she was asked, "What is the twenty-third root of this 201-digit number?"—bigger than millions of billions of trillions—she became the first person to ever beat a computer's calculations, and she did so in front of a live audience.

After that challenge Shakuntala was often called a "human computer"—a title she hated. She believed that the human mind had far more capabilities than computers, and she never wanted to feel limited in what she could achieve.

In a time when women were not given as many opportunities as men, Shakuntala proved herself to be a brilliant mathematician, a successful performer, and an independent woman. She was one of the first women to push forward in the field of mathematics, despite people insisting that women weren't smart enough or strong enough to participate in a field dominated by men.

Shakuntala wrote many mathematical books, including one called *Figuring: The Joy of Numbers*. She also became a famous astrologer, calculating people's horoscopes and star charts.

When people asked Shakuntala what her next challenge would be, Shakuntala always laughed and replied, "Nobody challenges me. I challenge myself."

Author's Note

In India, math is considered a highly important subject to study in school. For example, my family revered Srinivasa Ramanujan, a South Indian mathematician from our village in Tamil Nadu, who solved math problems that were considered insolvable.

So when I learned about Shakuntala Devi and her extraordinary mathematical talents, I was fascinated! How had such an interesting woman been hidden in history? As an early childhood educator, I knew that children don't typically begin to learn how to multiply and divide simple numbers until they are nine years old. And yet Shakuntala was able to multiply and divide complex numbers of ten digits or more without even a piece of paper when she was just three years old (something most adults can't do)!

Shakuntala's mathematical abilities were evidence of how a human can outperform technology—something that had never been seen before. She is an inspiration to generations of young women who continue to advance the opportunities for girls in STEM.

I see Shakuntala's story as a reminder that children have innate talents, and as unusual as those talents may be, they are a gift. By encouraging and uplifting those talents, we can help children to develop not only their skills but a healthy sense of self-esteem. And we can all strive for a sense of pride and confidence as strong as Shakuntala's!

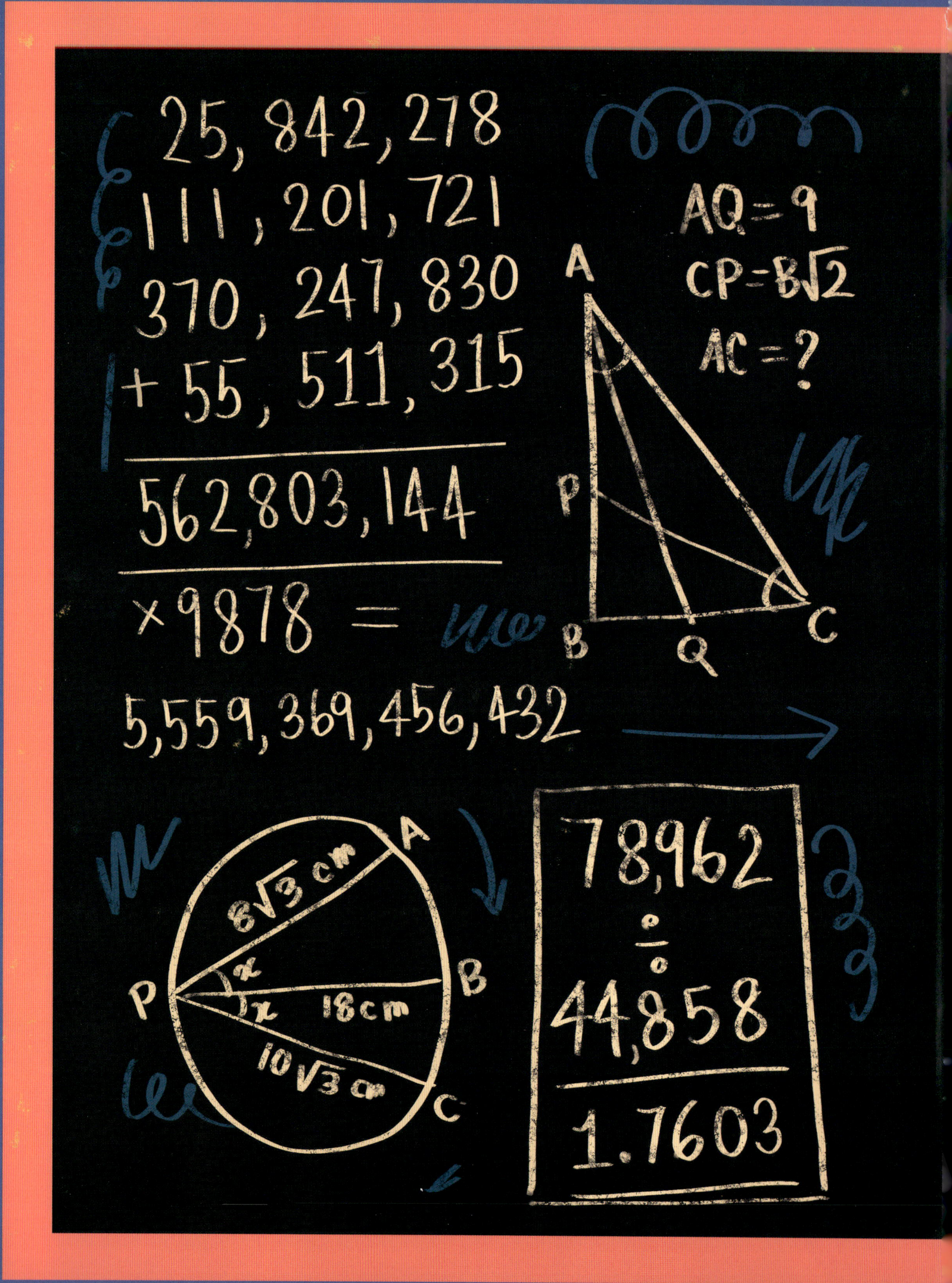
25,842,278
111,201,721
370,247,830
+ 55,511,315
562,803,144
×9878 =
5,559,369,456,432
AQ = 9
CP = 8√2
AC = ?
A
P
B
Q
C
8√3 cm
18 cm
10√3 cm
x
x
78,962
÷
44,858
1.7603